S0-AVQ-974

Published in 2015 by The Rosen Publishing Group, Inc.
29 East 21st Street, New York, NY 10010

Copyright © 2015 Weldon Owen Pty Ltd. Originally published in 2011 by Discovery Communications, LLC

Original copyright © 2011 Discovery Communications, LLC. Discovery Education™ and the Discovery Education logo are trademarks of Discovery Communications, LLC, used under license. All rights reserved.

All rights reserved. No part of this book may be reproduced in any form without permission in writing from the publisher, except by a reviewer.

Photo Credits: **KEY** t=top; tl=top left; tr=top right; cl=center left; c=center; bl=bottom left; bc=bottom center; br=bottom right; bg=background

GI = Getty Images; PDCD = PhotoDisc; SH = Shutterstock; TPL = photolibrary.com

6bg SH; **13**tl PDCD; **16**bl, br, c PDCD; **17**bc, tr PDCD; **23**c, t GI; **26**br, c PDCD; **29**bl, cl, cl, tl, tl SH; bl TPL

WELDON OWEN PTY LTD
Managing Director: Kay Scarlett
Creative Director: Sue Burk
Publisher: Helen Bateman
Senior Vice President, International Sales: Stuart Laurence
Vice President Sales North America: Ellen Towell
Administration Manager, International Sales: Kristine Ravn

Library of Congress Cataloging-in-Publication Data

Coupe, Robert, author.
 Predators and prey : battle for survival / by Robert Coupe.
    pages cm. — (Discovery education. Animals)
 Includes index.
 ISBN 978-1-4777-6956-0 (library binding) — ISBN 978-1-4777-6957-7 (pbk.) — ISBN 978-1-4777-6958-4 (6-pack)
 1. Predation (Biology)—Juvenile literature. 2. Predatory animals—Juvenile literature. 3. Animal behavior—Juvenile literature. I. Title.
 QL758.C75 2015
 591.5'3—dc23

2013047545

Manufactured in the United States of America

CPSIA Compliance Information: Batch #WS14PK3: For Further Information contact Rosen Publishing, New York, New York at 1-800-237-9932

PREDATORS AND PREY
BATTLE FOR SURVIVAL

ROBERT COUPE

New York

Contents

Food Chains....................................6

Predators from Long Ago 8

Self-Defense 10

Teeth 12

Talons and Claws 14

Poison......................................16

Moving Swiftly 18

Working Together 20

Blending In 22

Setting Traps 24

Big and Strong 26

Record Breakers............................28

Glossary....................................30

Index 32

Websites....................................32

Food Chains

All animals, plants, and bacteria are part of what we call a "food chain." For many animals in the wild, life is a constant battle for survival. They are always on the lookout for, hiding from, or trying to escape from animals that might kill them for food.

At the top of every food chain are animals, such as lions or tigers, that have no or very few predators. These top predators feed on animals farther down the food chain. Animals in the middle of the food chain hunt and eat animals that are lower down. Most animals, then, both feed on other animals and are themselves hunted for food. A few kinds of animals eat only plants. These are called herbivores.

MARINE FOOD CHAIN

Different food chains exist in different habitats. In a marine environment, large sharks are top predators. They have little to fear from other sea creatures. Bull sharks are among the most feared marine predators. Their large, triangular teeth can slice through the flesh of large fish, dolphins, sea turtles, and even other sharks. The large prey that they catch feed on smaller sea creatures, such as shrimp, which, in turn, eat small marine plants.

Shrimp

Large fish species

Bull shark

Short-toed eagle

Grass snake

Harvest mouse

Airborne predator

Eagles are top predators that hunt
from the air, but swoop earthward
to catch their prey. This short-toed
eagle has seized a grass snake.
The snake's diet includes small
mammals such as harvest mice,
which feed on seeds and insects.

Predators from Long Ago

Dinosaurs lived on Earth for about 160 million years, until they died out about 65 million years ago. We know a lot about these reptiles and how they lived, mainly from fossil remains of their skeletons and footprints that have been found in ancient rocks. Some dinosaurs were huge; some were small. Some moved swiftly; others slowly and clumsily. Most were meat-eaters; many fed only on plant matter.

Dinosaurs preyed on each other and on a range of other animals that shared their world. They used sharp teeth, powerful jaws, large claws, speed, and sheer size to catch and subdue their prey. They also used these, and other features such as spikes and heavy plates, to defend themselves against attack.

OTHER ANCIENT PREDATORS

For much of their time on Earth, dinosaurs were the most powerful animals. They shared the world with a range of other animals. Some hunted on the ground; others roamed the seas; a few could fly.

Utahraptor
This dinosaur had sawlike teeth and claws like sickles. It probably had feathers on its front limbs.

Cynognathus
Like today's wolves, *Cynognathus* hunted in packs. They were about the size of a wolf and had similar teeth.

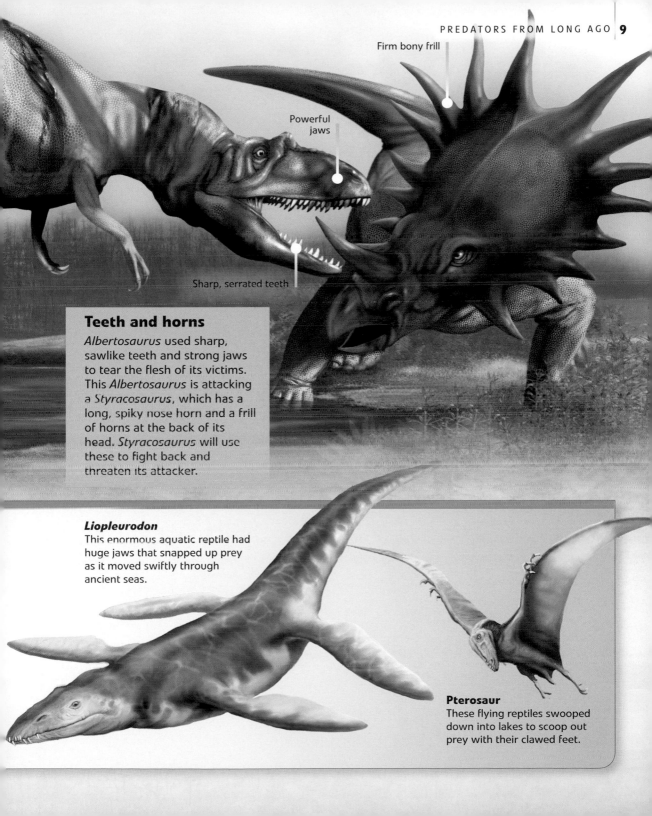

Firm bony frill

Powerful
jaws

Sharp, serrated teeth

Teeth and horns

Albertosaurus used sharp,
sawlike teeth and strong jaws
to tear the flesh of its victims.
This *Albertosaurus* is attacking
a *Styracosaurus*, which has a
long, spiky nose horn and a frill
of horns at the back of its
head. *Styracosaurus* will use
these to fight back and
threaten its attacker.

Liopleurodon
This enormous aquatic reptile had
huge jaws that snapped up prey
as it moved swiftly through
ancient seas.

Pterosaur
These flying reptiles swooped
down into lakes to scoop out
prey with their clawed feet.

Self-Defense

Many animals that are attacked by predators are able to fight back. Some can move fast enough to flee their attackers. Ostriches, for example, can usually outrun predators. But a cornered ostrich can kill its attacker with a powerful kick. Other animals have claws, horns, or hooves that they can use to injure or kill an attacker.

Nature has provided many species with inbuilt systems of self-defense. Some threatened animals change color or shape so that they blend with their background or can no longer be recognized. Some others have bright colors that warn predators of deadly poisons in their bodies. Others have sharp spikes that will deter a likely attacker.

That's Amazing!
Some lizards can escape a predator by shedding their tail. The wriggling tail distracts the attacker, giving the lizard time to escape. The tail later grows back.

Kicking back
Zebras have little defense against lions, but can fend off smaller attacking animals, such as African hunting dogs, with powerful kicks from their sharp-edged back hooves.

Sharp reply
When danger threatens the echidna, an Australian native, it rolls itself into a ball and exposes its needle-sharp spines.

Frightening frog
Poison dart frogs come in a range of colors. These colors warn predators that these frogs are dangerous to eat.

Smelly snakes
If they sense danger, some snakes can roll over and give off a foul smell. This deters any would-be predator.

Shell shelter
Hermit crabs crawl into empty shells so that octopuses and other predators will not find them. The shells provide camouflage for the crabs.

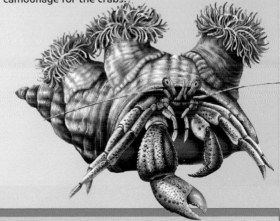

SKUNK SMELL

A skunk will try to run from an attacker. If it cannot, it will squirt out a nasty-smelling liquid from beneath its tail.

Raises tail as a warning.

Stands up on its front legs.

Squirts out liquid from under tail.

Skunk stares at attacker.

A varied mouthful

Like all cats, tigers have long canine teeth at the front of their mouth. They use these to seize and kill their prey. Between these canine teeth, at the top and bottom of the mouth, is a line of smaller incisor teeth.

Cats' jaws can move only up and down, so their teeth cannot move from side to side to grind up food.

Teeth

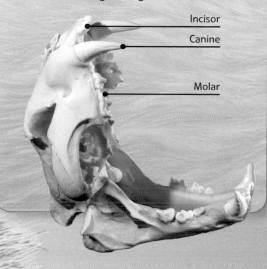

TEETH AND JAWS

Tigers have powerful jaw muscles. While the canine teeth grab and kill prey, the incisors keep a grip on it. The molars, farther back in the mouth, act like scissors, cutting through the flesh.

Incisor

Canine

Molar

Teeth are one of the main weapons of many hunters in the animal world, especially those that catch large prey. Both on land and in water, predators take their victims with their teeth, or use their teeth to kill and then tear their prey's flesh.

Predators' teeth vary greatly in size, shape, strength, and sharpness. And just as people have several different kinds of teeth in their mouth, so do most predatory animals. Different teeth do different jobs: they can seize prey, stab it, crush it, grip it firmly, and grind it up. Some sharks have teeth with sawlike serrated edges that help them slice up the tough skin of seals and other marine mammals.

Lizard teeth
The Komodo dragon, the world's largest lizard, bites into its large prey with fanglike teeth. The victim is poisoned by the deadly bacteria in the dragon's mouth and dies slowly. The dragon then tears up its prey with its teeth.

Talons and Claws

Many animal hunters have long, sharp, nail-like claws at the end of their toes. Some animals use these as weapons if they get into a fight. Big cats use them to grasp prey they pounce on before attacking with their teeth. Most cats can pull their claws back into their paws when they are at rest.

Birds of prey, such as eagles, hawks, and owls, have claws that are usually called talons. They use them to grip, then carry away the prey that they seize on the ground as they swoop down from on high.

CLAWS AND TEETH

A polar bear will lie in wait on the polar ice and then suddenly seize a seal that comes up for air in its long, sharp, strong claws. It then grabs the seal in its teeth and tears off the flesh.

Sharp and rounded
Different kinds of bears use their claws for digging out, slashing, or spearing their prey.

Digging deep
A bear's claw can grow as long as 4 inches (10 cm).

Fingers and toes
At the end of each of its front and back paws, a polar bear has five claws, just like the nails on human fingers and toes.

Crab claws
At the end of each of its front legs, a crab has a pair of claws, or pincers, that come together to grab and crush prey, such as shrimp.

Wide span
A polar bear's massive paws measure 12 inches (30 cm) across.

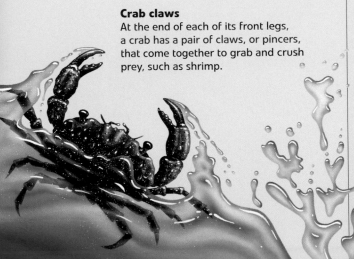

Viselike grip

With its super-sharp eyesight, an owl locates its prey. It then dives and grabs its victim in its sharp, curved talons.

Holding on
The talons lock into position to hold the prey in place.

Talons spread wide
Before grabbing its victim, an owl spreads its talons out wide.

Poison

Poison is another powerful weapon that many animals use in catching and killing their prey. Some animals, such as certain species of frogs, have poisons in their bodies that help protect them from an attack. Others have sharp fangs that they use to inject poison from within their bodies into animals they bite. Some creatures have deadly stings in their tail or tentacles that stun or kill prey that get caught in them.

Rearing up to strike

There are many kinds of venomous spiders. Funnel-web spiders in Australia inject powerful venom into their prey and any human who may get bitten. Perhaps the most deadly of all is the Brazilian wandering spider's venom. These spiders rear up before plunging their fangs into their victims.

Brazilian wandering spider

Lethal lizard

The Gila monster lives in desert areas of the southwestern US. When it bites its prey, poisonous saliva from the Gila monster's mouth flows into the wound and slowly kills the prey.

A sting in the tail

Like crabs, scorpions have pincer claws. They catch and hold their prey in these claws. To ensure the prey does not escape, a scorpion may then poison it with the venomous sting in its tail.

Stinging tentacles

The Portuguese man-of-war dangles its long, stinging tentacles underneath as it floats through the ocean. The tentacles trap and poison small fish and shrimp. Some kinds of fish can swim unharmed through the tentacles.

Silent hunter

Rattlesnakes are desert animals. They ambush their prey by hiding under dried leaves and using their pit organs (infra-red-sensitive organs located between the eye and the nose) to "smell out" their victims. When a rattlesnake strikes out at a mouse or a rat, it injects poison through its needle-like fangs. When its mouth is closed, the long fangs fold back along the top of the mouth.

Moving Swiftly

An ability to move swiftly keeps a lot of animals alive. It allows many predators to chase and catch the prey they need in order to survive. It helps others to escape from an attacker by outrunning it. Whether in the air, on the ground, in the oceans, or even high up in forest trees, speed can be a vital survival tool. Some hunters need to combine speed with stealth. Tigers can run fast but not for long. They need to sneak up on their prey and then spring on it for a quick kill.

Shooting tongue

Chameleons do not break any sprint records, but their long tongue moves at great speed. Most of these lizards live in rain forest trees on the island of Madagascar. When a chameleon spots an insect, its tongue shoots out rapidly and traps the prey on its sticky end.

A chameleon's tongue reaches its target.

World record
The peregrine falcon is the fastest-moving animal on Earth. It preys on other birds and, when it dives to catch one, it can reach speeds of 200 miles (322 km) per hour.

Faster and faster
Cheetahs are the world's fastest land animals. When chasing prey, they can reach 45 miles (72 km) per hour in just 2.5 seconds. From there they can accelerate farther to 64 miles (103 km) per hour.

Water speed record
Sailfish are probably the world's fastest-moving fish. They are common in warmer oceans but, in recent times, fishing has reduced their numbers. When hunting prey, they can hurtle through the water at 68 miles (110 km) per hour. They charge at schools of smaller fish or squid, and ram them with their long, spiky bill.

The sailfish's spiny sail goes up only as a warning sign, when this fish feels that it is in danger.

Hunting moose

A moose weighs more than 12 times as much as a wolf, but has no chance against a pack of six or eight hungry wolves. Each wolf attacks a different part of the moose's body.

Working Together

Some animals live and hunt alone. Others live and hunt in communities, teams, or packs. Smaller animals that hunt in groups have a much better chance of catching and killing much larger prey. You would never see a single ant take on a grasshopper, but you may have seen a swarm of them attacking a much larger insect.

Wolves and other wild dogs are great pack hunters. They use howls and other sounds, as well as scents and body movements, to send signals to each other. Being in a group can also make life safer, as any one individual is less likely to be caught. Adult white rhinos in Africa will form a circle around a young rhino to protect it from attack by hyenas or other predators.

DELIVERING DINNER

Groups, or pods, of dolphins cooperate to catch a meal. When they find a school of fish, one dolphin swims in circles around it and herds it, like a sheepdog with sheep, toward the other dolphins, which then move in to eat.

Warning signs
The markings on the wings of this moth look like eyes staring up. They can trick a predator into thinking it is too dangerous to attack.

Blending In

Many animals have colors or patterns that allow them to blend in to their environment. This makes it hard for predators to see them. These colors and markings are called camouflage.

Camouflage helps many animals to hide from would-be attackers, and also helps many predators to stalk their prey and get close to it without being detected. A tiger's stripes, for example, make it hard to see the animal in long grasses as it sneaks up on its prey. The stripes on one zebra blend in with all the other zebra stripes, so that a herd looks like one large mass to a predator. This makes it difficult for the predator, usually a lion, to pick out and go after any individual zebra.

HIDDEN PREDATORS

Camouflage helps predators to hunt, and prey to hide. Here are examples of predators whose camouflage helps them sneak up on their prey and then launch a surprise attack.

Flaps down
The Malaysian flying gecko flattens the flaps on the sides of its body as it sneaks up on prey. Its markings blend with its surroundings and, being flat, it does not create a shadow.

Spots and pebbles
The spotted pattern on the leopard flounder's body blends with the pebbly ocean floor as it hunts for crabs.

Snowy coat
A seal coming up for air may not see a polar bear that is camouflaged against the snow and ice until it is too late.

Different pigments

A chameleon has see-through skin with layers of yellow, red, brown, and blue pigments. These combine to produce different and changing combinations of colors.

Pigment cells

Warning

Quick change

Chameleons are lizards that can change color quickly. These color changes are used to communicate. They can also frighten off snakes or other predators.

Camouflage

Attack or defense?

The same patch of skin changes color, as dark brown cells either reduce in size to reveal other-colored cells, or expand to cover them.

Setting Traps

Most predators go searching for their prey. Some, however, build traps and just wait for their victims to fall or fly into them. Spiders are the great trap-builders. Most of them spin large, sticky webs that are woven from silky thread from within their bodies. These webs hang in the air and trap flies and other insects that fly into them and become entangled. A trapdoor spider lurks underground in a camouflaged den that its victims cannot see until it is too late to escape.

That's Amazing!

Some plants catch insects and small animals. These "meat-eating" plants grow around the world, mainly in rain forests and other wet environments, such as bogs and ponds.

Setting the trap
A trapdoor spider builds an underground burrow with a hinged, camouflaged lid. It then lies in wait.

Catching prey
When it senses an insect approaching, the spider springs up through the lid and seizes its prey.

VENUS FLY TRAP

Plants cannot move, but some of them are traps that catch insects. The Venus fly trap is one of these. It has two open leaves that snap shut.

1 Attracted
Bright colors and sweet smells attract a fly to the Venus fly trap.

2 Entering
The fly enters the leaves of the plant and touches hairs on the inside.

3 Trapped
The leaves snap shut, and the plant begins to digest the fly.

Writing spider

This spider spins a thick X-shaped pattern in its web to make it stronger. This pattern looks like a child's scribble, which is how these spiders got their name.

Wrapped up
The spider injects the fly with poison, then wraps it up in silk to be eaten later.

Getting stuck
A fly flies into the web and gets stuck.

Big and Strong

Size and strength are certainly useful in the wild. Large, powerful predators, such as lions and tigers, have an advantage over smaller hunters. Not only are they able to catch larger prey, their size and strength protect them from attacks from other hunters.

Sheer strength is the main weapon that some species can use in their pursuit of prey. Most snakes use venom to subdue their victims, but large constrictor snakes, such as pythons and boas, use the strength of their muscles as they wrap their bodies around their prey and crush it to death. Chimpanzees also employ brute strength to kill smaller monkeys by dashing them against the ground.

Crocodile meat

Burmese pythons live in forests, woodlands, and swamps of Southeast Asia. They grow up to 20 feet (6 m) long, and feed on mammals, birds, and reptiles. This one has just crushed to death a young Siamese crocodile, and is beginning to swallow it head first.

Did You Know?

Humans have long hunted and killed elephants for the ivory in their tusks. This hunting is now illegal, but still continues in some places.

Record breakers

Sperm whales are the world's largest toothed predators. They have larger teeth and larger brains than any other animal, and dive deeper in the ocean than any other marine mammal to catch prey as large as giant squid. They can grow to 60 feet (18 m) long and weigh up to 50 tons (45 t).

Safe from predators

Elephants are not predators. They feed on a wide variety of grasses and plant material. They are the world's largest land animals. Their sheer size and strength protect them from attack from other animals.

Largest animal
The blue whale is the world's largest animal.

Tiny prey
The blue whale eats mainly tiny sea creatures called krill.

Quick motion
A trap-jaw ant's jaws and a mantis shrimp's front legs move much faster than our eyes can see.

Mantis shrimp

Record Breakers

Animals hold all sorts of amazing world records. Some insects break records for traveling in huge numbers. Swarms of up to 20 million ants can bring death and destruction to animals in their path. Killer bees, first bred in Brazil in the 1950s, sometimes travel in great swarms, biting and killing people and animals that get in their way.

Trap-jaw ant

African driver ants

Dangerous swarms
African driver ants move slowly across the country, so people have time to get away from them. Killer bees are more difficult to avoid.

Killer bees

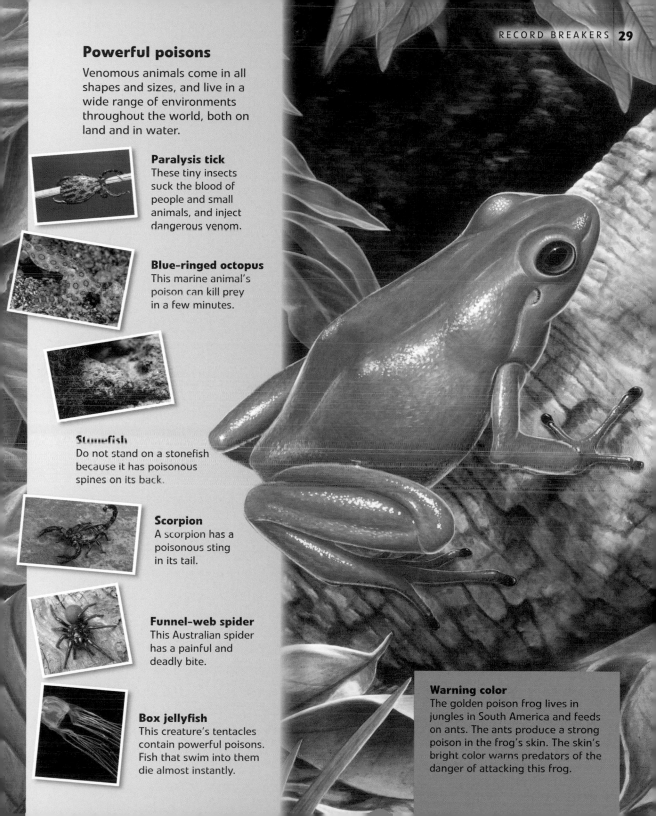

Powerful poisons

Venomous animals come in all shapes and sizes, and live in a wide range of environments throughout the world, both on land and in water.

Paralysis tick
These tiny insects suck the blood of people and small animals, and inject dangerous venom.

Blue-ringed octopus
This marine animal's poison can kill prey in a few minutes.

Stonefish
Do not stand on a stonefish because it has poisonous spines on its back.

Scorpion
A scorpion has a poisonous sting in its tail.

Funnel-web spider
This Australian spider has a painful and deadly bite.

Box jellyfish
This creature's tentacles contain powerful poisons. Fish that swim into them die almost instantly.

Warning color
The golden poison frog lives in jungles in South America and feeds on ants. The ants produce a strong poison in the frog's skin. The skin's bright color warns predators of the danger of attacking this frog.

Glossary

aquatic (uh-KWAH-tik)
An aquatic animal is one that lives entirely or partly in water. Frogs are aquatic animals.

bacteria (bak-TIR-ee-uh)
A group of microscopic, single-celled organisms. Each cell is called a bacterium. Some bacteria help human and animal bodies to work normally and keep healthy. Other bacteria are harmful and cause disease.

birds of prey
(BURDZ UV PRAY) Birds, such as eagles, hawks, falcons, and owls, that hunt and eat other birds and animals. They often swoop down on their victims and catch and carry them away in their sharp claws, or talons.

camouflage
(KA-muh-flahj) Colors and patterns on an animal's body that help it to blend in with its surroundings and make it hard to see. Camouflage helps to protect many animals from attack.

canine teeth
(KAY-nyn TEETH) Long, pointed teeth near the front of a mammal's mouth. Many predators use canine teeth to catch and kill their prey.

dinosaurs
(DY-nuh-sawrz) A group of reptiles that were the dominant animals on Earth for about 160 million years until they died out about 65 million years ago.

environment
(en-VY-ern-ment)
The natural surroundings in which an animal usually lives.

fangs (FANGZ) Sharp teeth that have a groove or canal that can inject poison into an animal that is bitten. Some snakes and spiders have poisonous fangs.

food chain (FOOD CHAYN)
A pattern of feeding in which animals at higher levels in the food chain feed on those lower down the chain. Animals at the top of a food chain are those that are not hunted by other animals. If some animals in a food chain die out, animals higher up the food chain may be endangered.

habitat (HA-buh-tat)
The kind of environment in which an animal lives in the wild, such as a forest, a desert, a river, or a part of the ocean. A habitat provides the animal with the food and shelter, and other things that it needs to survive.

herbivore
(ER-buh-vor)
An animal that eats only, or mainly, plant material and does not hunt other animals. Many herbivores are hunted by other animals.

herd (HURD)
A group of animals, such as cattle or buffalo, that come together and move around together.

incisor teeth
(in-SY-zur TEETH)
The front teeth, in the upper and lower jaws of mammals, between the longer and sharper canine teeth. Incisor teeth are used for cutting through food.

krill (KRIL)
Tiny, shrimplike marine creatures that live in huge numbers in the cold waters around the North and South poles. Krill are the main food of some large whales.

mammal
(MA-mul)
A warm-blooded animal that has a backbone and hair on its body, and that feeds its young on milk from its body. Humans are mammals. So are dogs, cats, seals, and whales.

marine (muh-REEN)
A marine animal lives mainly or wholly in the world's seas and oceans. Most fish are marine animals. So are whales and seals.

molars
(MOH-lurz) Low, flat teeth at the top and bottom of a mammal's mouth, behind the canine teeth. Molars are used for chewing and grinding up food.

pack (PAK) A group of animals, such as dogs or wolves, that hunt together.

pigment cells
(PIG-ment SELZ) The cells that control the skin and eye color of animals.

predator
(PREH-duh-ter) An animal that hunts and kills other animals for food.

prey (PRAY) An animal that is hunted and killed for food by other animals.

reptile (REP-tyl)
A cold-blooded animal that has a backbone, breathes air, and has scaly skin. Most reptiles lay eggs. Crocodiles, lizards, snakes, and turtles are all reptiles.

saliva (suh-LY-vuh) The watery liquid that glands produce in the mouth of humans and other animals. Also called spittle, saliva helps animals to digest their food.

school (SKOOL) The name for a large number of fish that come together and swim in a group.

serrated (ser-AYT-ed)
Rough and sharp at the edges, similar to the edge of a saw.

species (SPEE-sheez)
Groups of animals that are similar in appearance and have many other features in common. Members of the same species mate to produce offspring.

stealth (STELTH)
The ability to creep up close to prey without being seen or heard.

survival (sur-VY-val)
Staying alive by avoiding being hunted and killed by other animals.

swarm
(SWORM)
Insects, such as bees, ants, and locusts, sometimes move across vast distances in huge numbers, creating great damage. When they move in this way, it is called a swarm.

talons
(TA-lunz) The long, sharp claws that birds of prey use to grab and carry away their prey.

tentacles
(TEN-tih-kulz)
The long, thin, outer parts on some marine creatures, such as octopuses and jellyfish. Tentacles feel and grab hold of prey and, in some cases, inject poison into it.

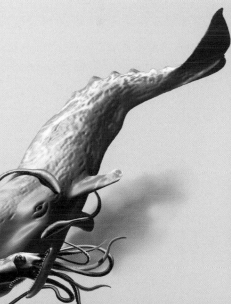

Index

A

Albertosaurus 9
ants 21, 28

B

bees 28
bull sharks 6

C

chameleons 18, 23
cheetahs 19
crabs 11, 14, 22
crocodiles 26

D

dinosaurs 8, 9
dolphins 21

E

eagles 7, 14
elephants 26, 27

F

fish 19, 21, 22, 29
food chains 6, 7
frogs 11, 16, 29

G

Gila monsters 16
grass snakes 7

H

harvest mice 7
hawks 14

I

insects 21, 22, 24,
    25, 28, 29

J

jaws 12, 13

K

Komodo dragons 13
krill 28

L

leopard flounder 22
lions 22, 26
lizards 10, 13, 16, 22, 23

M

Madagascar 18
Malaysian flying geckos 22
moose 20
moths 22

O

ostriches 10
owls 14, 15

P

paralysis ticks 29
peregrene falcons 19
polar bears 14, 22
Portuguese man-of-wars 17
pterosaurs 9

R

rattlesnakes 17

S

sailfish 19
scorpions 16, 29
seals 22
sea turtles 6
sharks 6
short-toed eagles 7
skunks 11
snakes 7, 11, 17, 26
sperm whales 26
spiders 16, 24,
    25, 29
squid 26
Styracosaurus 9

T

teeth 12, 13
tigers 6, 12, 18, 22, 26
trapdoor spiders 24

V

Venus fly traps 24

W

whales 26, 28
white rhinoceroces 21
writing spiders 25

Z

zebras 10, 22

Websites

Due to the changing nature of Internet links, PowerKids Press has developed an online list of websites related to the subject of this book. This site is updated regularly. Please use this link to access the list:

www.powerkidslinks.com/disc/surv/

Dolores County Children's Branch
PO Box 578
Dove Creek, CO 81324-0578